Community Helpers

Animal Control Officers

by Erika S. Manley

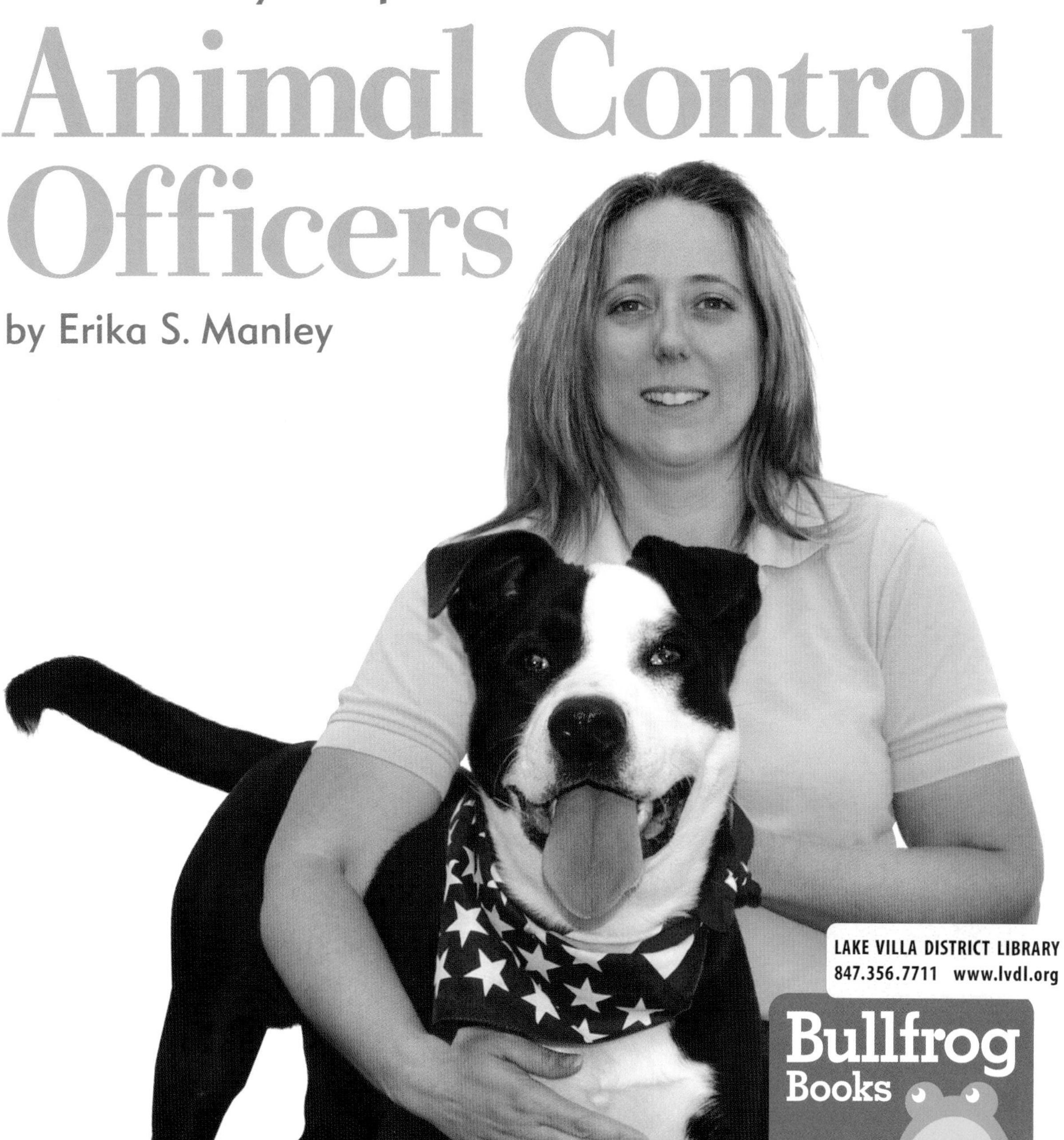

Bullfrog Books

Ideas for Parents and Teachers

Bullfrog Books let children practice reading informational text at the earliest reading levels. Repetition, familiar words, and photo labels support early readers.

Before Reading

- Discuss the cover photo. What does it tell them?
- Look at the picture glossary together. Read and discuss the words.

Read the Book

- "Walk" through the book and look at the photos. Let the child ask questions. Point out the photo labels.
- Read the book to the child, or have him or her read independently.

After Reading

- Prompt the child to think more. Ask: Have you heard about animal control officers before reading this book? What more would you like to learn about them after reading it?

Bullfrog Books are published by Jump!
5357 Penn Avenue South
Minneapolis, MN 55419
www.jumplibrary.com

Library of Congress Cataloging-in-Publication Data

Names: Manley, Erika S., author.
Title: Animal control officers / by Erika S. Manley.
Description: Minneapolis, MN: Jump!, Inc., [2020]
Series: Community helpers | "Bullfrog Books are published by Jump!"
Audience: Ages 5–8. | Audience: K to grade 3.
Includes index.
Identifiers: LCCN 2018053668 (print)
LCCN 2018059920 (ebook)
ISBN 9781641288279 (ebook)
ISBN 9781641288255 (hardcover : alk. paper)
ISBN 9781641288262 (pbk.)
Subjects: LCSH: Animal welfare—Juvenile literature.
Animal specialists—Juvenile literature.
Classification: LCC SF80 (ebook)
LCC SF80 .M348 2020 (print) | DDC 636.08/32—dc23
LC record available at https://lccn.loc.gov/2018053668

Editor: Jenna Trnka
Design: Shoreline Publishing Group

Photo Credits: sirastock/iStock, cover (foreground); photosbyjim/iStock, cover (background); The Journal/Mary Stortstrom/AP Images, 1; Elisabeth Burrell/Dreamstime, 3; Paulus Rusyanto/Dreamstime, 4; ZUMA Press Inc./Alamy, 5, 14–15; Adam Eschbach/Idaho Press, 6–7; Lisa Eastman/Shutterstock, 8–9, 22tr; Jose Galemore/The Casper Star-Tribune/AP Images, 10; Hedgehog94/Shutterstock, 11, 23tr; Suzanne Kreiter/Boston Globe/Getty Images, 12–13, 23br; Willeecole/Dreamstime, 16; Jill Morgan/Alamy, 17, 22tl; FatCamera/iStock, 18–19, 23tl; Capital Gazette/Joshua McKerrow/AP Images, 20–21; Willeecole Photography/Shutterstock, 22bl; Fotyma/Shutterstock, 22br; Tom Wang/Dreamstime, 23bl; Creativa Images/Shutterstock, 24.

Printed in the United States of America at Corporate Graphics in North Mankato, Minnesota.

Table of Contents

Animal Helpers

Maya wants to be an animal control officer.

What do they do?

They help animals.

Mia finds this lost dog.

She puts a leash on it.

leash

catch pole

This dog tries to run.

Mike catches it.

He uses a catch pole.

Where do they take a lost animal?

A shelter.

The owner can pick it up!

glove

This wild animal
is on the loose!

Who do we call?

Animal control!

Tina wears
a glove.

Oh, no!

A cat fell.

Tracy will get it!

She can use a net.

She gets a kennel.

She will bring
the cat to the vet.

We can adopt the cat!
We will care for it.

ANIMAL CONTROL
COUNTY POLICE
OFFICER

Animal control officers do good work!

Tools for the Job

animal control vehicle
A special vehicle with several enclosed stalls to keep animals safe.

catch pole
A pole with a loop at one end that fits around an animal's neck.

kennel
A cage used to keep an animal contained.

leash
A rope that attaches to an animal's collar.

Picture Glossary

adopt
To take in an animal to live in your home.

shelter
A temporary home for a rescued animal.

vet
Short for veterinarian, a vet is a doctor who cares for animals.

wild animal
An animal that is not tame and lives in the wild without any help from people.

Index

To Learn More

Finding more information is as easy as 1, 2, 3.

1. Go to www.factsurfer.com
2. Enter "animalcontrolofficers" into the search box.
3. Choose your book to see a list of websites.